Spring Food

Julie Murray

Abdo Kids Junior
is an Imprint of Abdo Kids
abdobooks.com

SEASONS: SPRING CHEER!

Abdo Kids

abdobooks.com

Published by Abdo Kids, a division of ABDO, P.O. Box 398166, Minneapolis, Minnesota 55439. Copyright © 2021 by Abdo Consulting Group, Inc. International copyrights reserved in all countries. No part of this book may be reproduced in any form without written permission from the publisher. Abdo Kids Junior™ is a trademark and logo of Abdo Kids.

Printed in the United States of America, North Mankato, Minnesota.

102020
012021

THIS BOOK CONTAINS RECYCLED MATERIALS

Photo Credits: AP Images, iStock, Shutterstock

Production Contributors: Teddy Borth, Jennie Forsberg, Grace Hansen

Design Contributors: Candice Keimig, Pakou Moua, Dorothy Toth

Library of Congress Control Number: 2020910601

Publisher's Cataloging-in-Publication Data

Names: Murray, Julie, author.

Title: Spring food / by Julie Murray

Description: Minneapolis, Minnesota : Abdo Kids, 2021 | Series: Seasons: spring cheer! | Includes online resources and index.

Identifiers: ISBN 9781098205874 (lib. bdg.) | ISBN 9781098206437 (ebook) | ISBN 9781098206710 (Read-to-Me ebook)

Subjects: LCSH: Spring--Juvenile literature. | Food--Juvenile literature. | Food supply--Seasonal variations--Juvenile literature. | Spring plants--Juvenile literature. | Seasons--Juvenile literature.

Classification: DDC 508.20--dc23

Table of Contents

Spring Food 4

More Spring Food . . .22

Glossary 23

Index 24

Abdo Kids Code 24

Spring Food

Spring is here! Fresh food is growing.

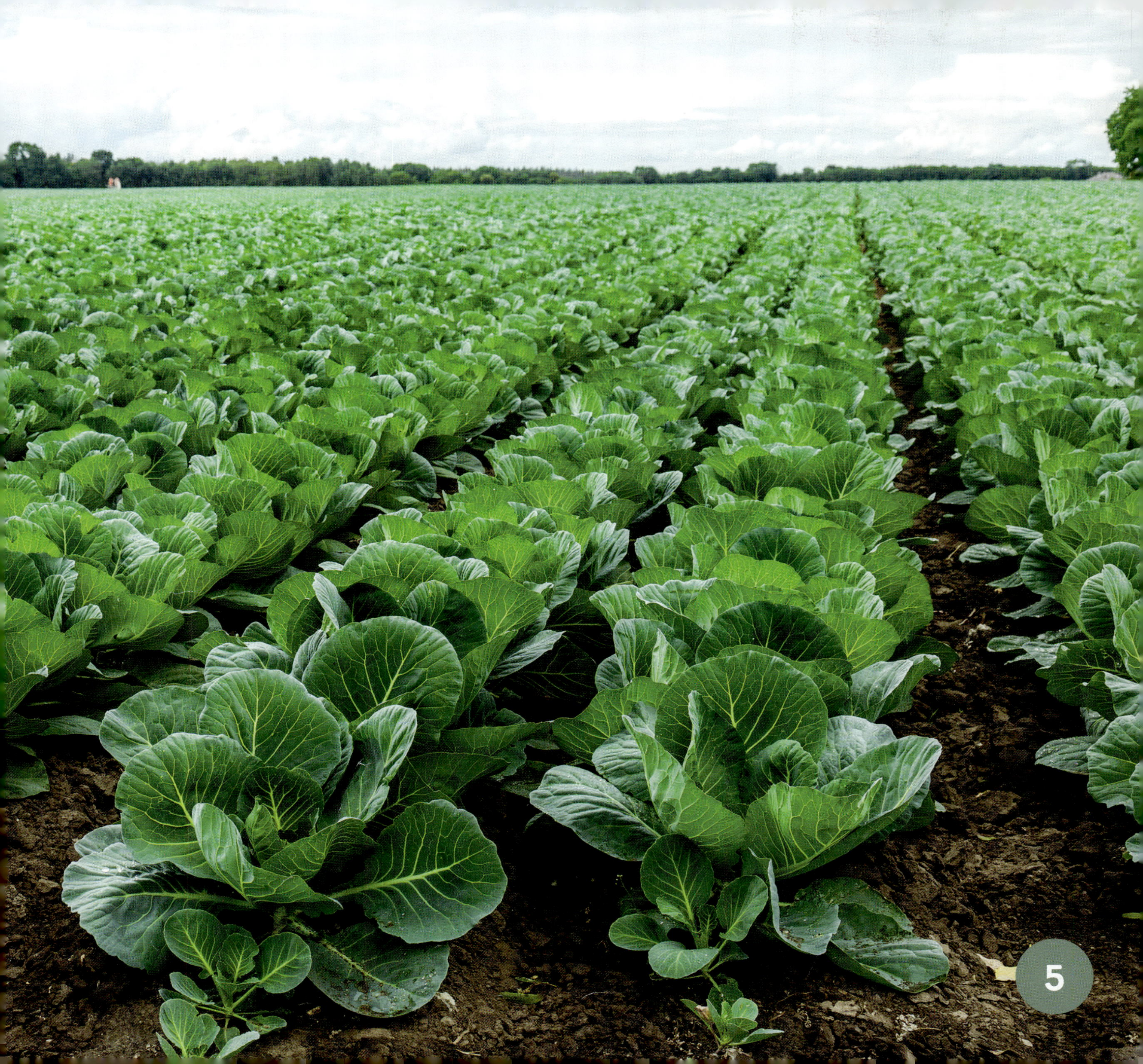

The strawberries are red.

Joe picks them.

7

Mom cuts the carrots.

They are yummy!

9

The grapefruit are **ripe**!

Leo sips the juice.

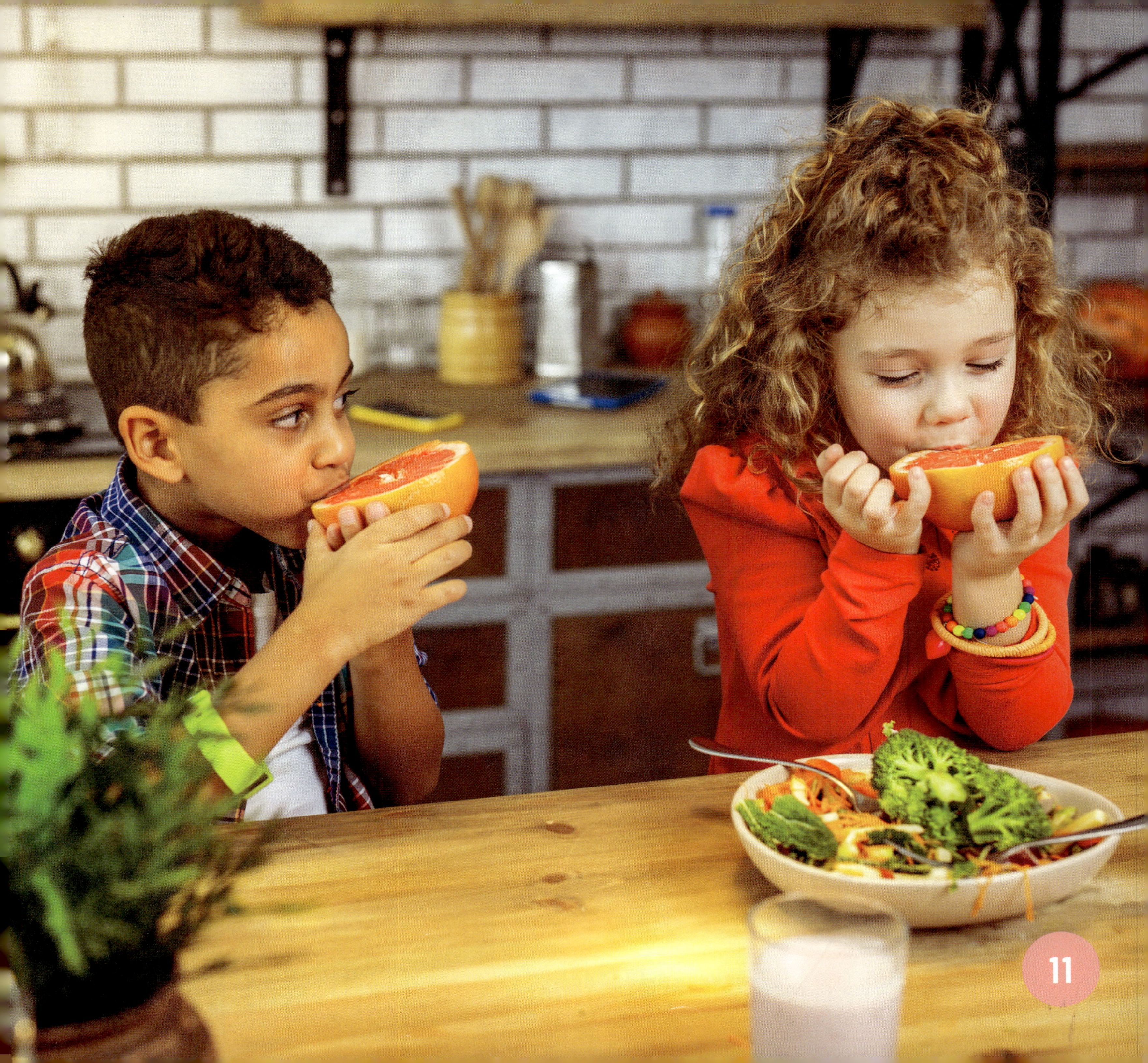

Zach loves asparagus. His mom **grills** some for dinner.

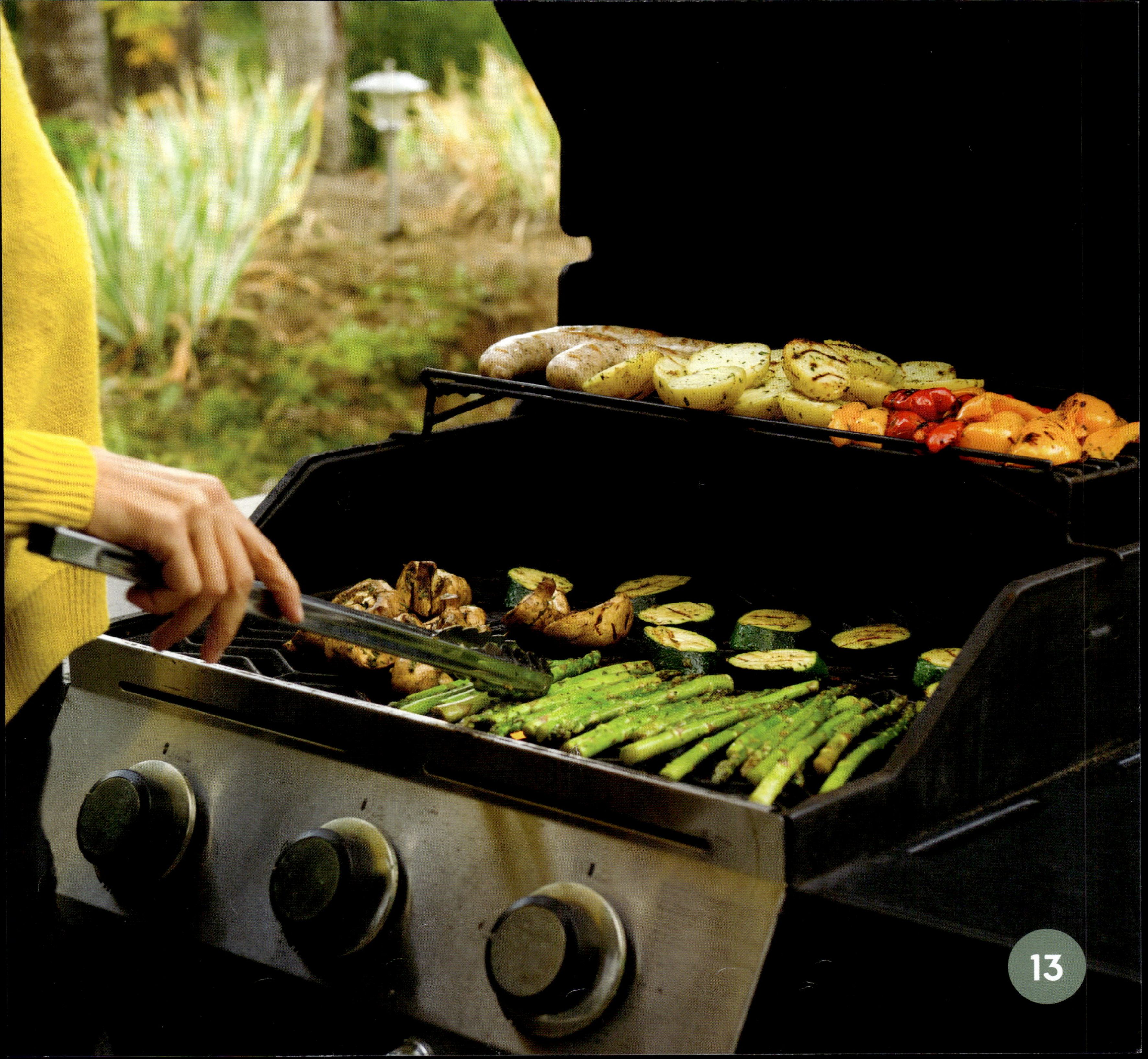

The tree is full of apricots.

Ivy loves to eat them!

15

Lou helps his dad. They **harvest** maple syrup.

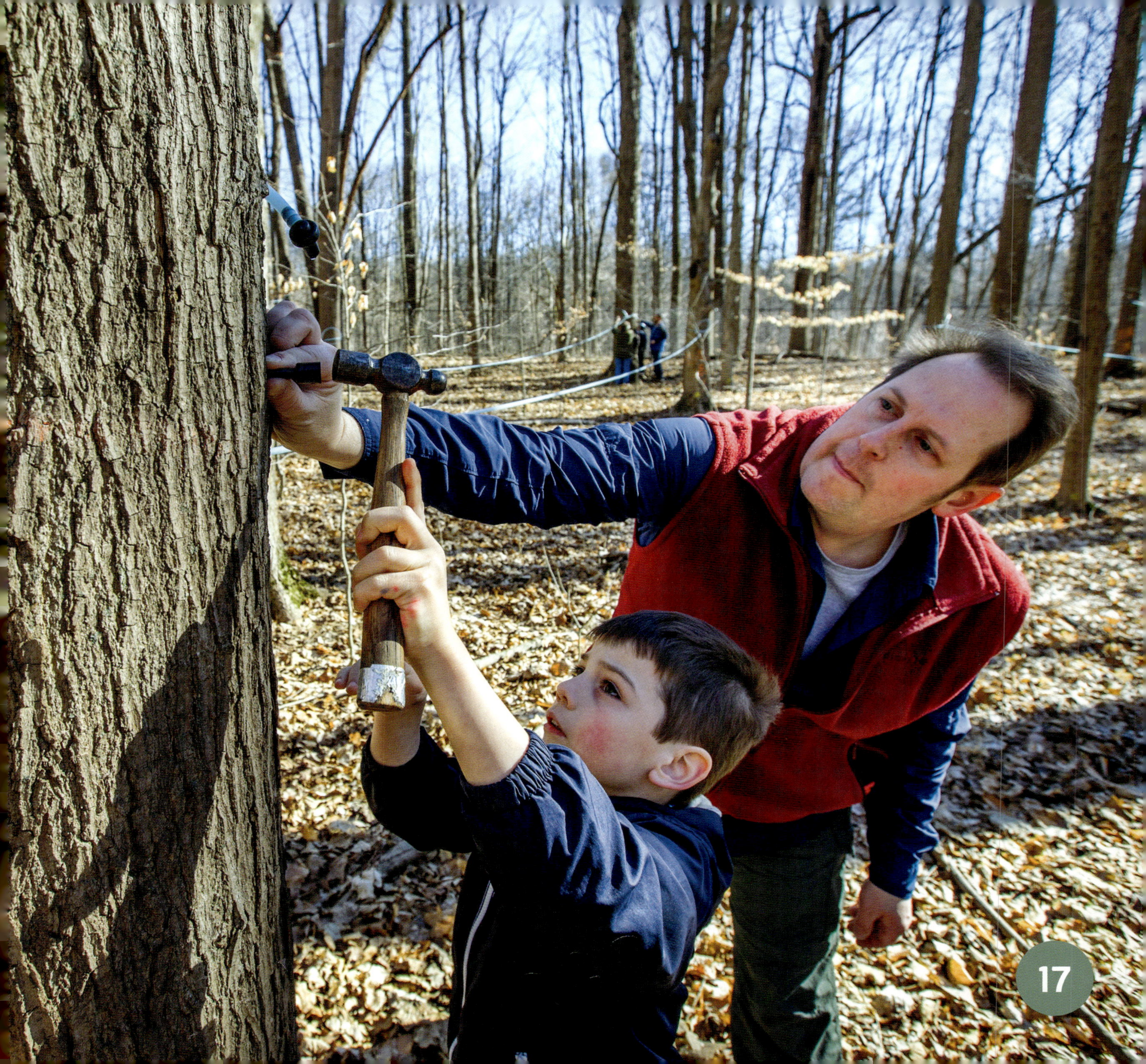

Chris picks the peas. The peas are bright green.

19

What spring food do you like?

21

More Spring Food

artichoke

kiwi

radish

rhubarb

22

Glossary

harvest
the gathering of food that is ready or done growing.

grill
to cook food on a rack of metal bars that sits over flames or high heat.

ripe
fully grown and ready for eating.

Index

apricots 14

asparagus 12

carrots 8

grapefruit 10

maple syrup 16

peas 18

strawberries 6

Abdo Kids ONLINE
FREE! ONLINE MULTIMEDIA RESOURCES

Visit **abdokids.com** to access crafts, games, videos, and more!

Use Abdo Kids code **SSK5874** or scan this QR code!